the dodo

Little but Fierce

By Joan Emerson

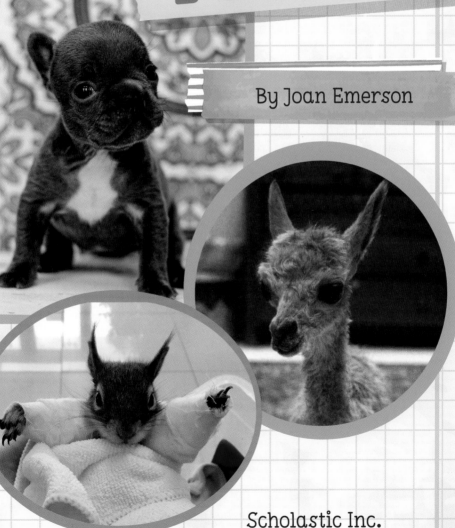

Scholastic Inc.

Photo credits: Cover: © Cara Smyth; title page (top): © Cara Smyth; title page (bottom): © Tayfun Demir; title page (right): © Cody, The Teeny Tiny Alpaca; pages 3-12: © Cara Smyth; pages 13-22: © Cody, The Teeny Tiny Alpaca; pages 23-31: © Tayfun Demir

All rights reserved. Published by Scholastic Inc., *Publishers since 1920.* SCHOLASTIC and associated logos are trademarks and/or registered trademarks of Scholastic Inc.

ISBN 978-1-338-62181-5

10 9 8 7 6 5 4 3 2 1   20 21 22 23 24

Printed in the U.S.A.   113

This edition first printing, February 2020

Cover design by Mercedes Padró
Interior design by Kay Petronio

VERA

the French Bulldog

When Vera the French bulldog was born, she could fit into a teacup!

But that wasn't the only thing different about Vera.

Vera also had a condition called cleft palate. That means that the roof of her mouth did not form properly.

Vera's condition made it hard to eat.
She needed some extra help to get healthy.

When Vera was just one hour old, a woman met the puppy and fell in love!

Vera's new owner knew this little puppy would be a lot of work. But she wanted to help.

Vera needed around-the-clock care.
Vera's owner began to feed the puppy with
a feeding tube every two hours.

After just four weeks, Vera was eating solid food.
She could drink out of a hamster bottle, too.

Soon Vera even started walking on her own feet. She fell down every now and then, but that was okay.

It was clear Vera wanted to grow up to be big and strong.

Her owner knew this silly puppy was here to stay!

Vera is still small. But she doesn't let that stop her.

In fact, Vera acts like she doesn't know how tiny she really is!

Vera plays with toys that are three times her size.

And she even became friends with the bigger cats and dogs who live in the house.

Vera just wants to play all day.
Everyone in her family is happy to jump in.
It only takes a little love to make a **BIG**
difference!

CODY

the Alpaca

Cody the little alpaca was born on a ranch in Colorado.

There were more than one hundred alpacas on the ranch. But Cody was special.

When Cody was born, she only weighed about six pounds.

That is a third of the size of most alpaca babies!

She was so small that she could not stand on her own.

But Cody's owner could tell this little alpaca was a fighter.

Cody needed some extra love and care.
So Cody moved into her owner's house.

This was hard for the family to get used to.

Cody eats hay, and she is pretty messy! But soon they figured it out.

Cody has lots of fun living inside.
She likes to cuddle and play with her toys.
She even gets to go upstairs for bedtime!

But Cody's small size still made life hard sometimes.

When she was young, Cody broke her leg twice. Sometimes she felt so weak, it was hard to stand up.

But Cody beat the odds with strength, a great family, and a little luck!

When Cody got bigger and stronger, she was able to go outside with the other alpacas.

At first, tiny Cody wasn't sure what to do with the bigger animals.

But after a while, they became her close friends

Now Cody explores the ranch during the day with her alpaca friends.

At night, she comes inside and spends time with her family.

Cody may be tiny, but her heart is BIG!

KARAMEL

the Squirrel

Karamel the little squirrel was hurt in a hunter's trap.

But one man heard Karamel's story and came to her rescue.

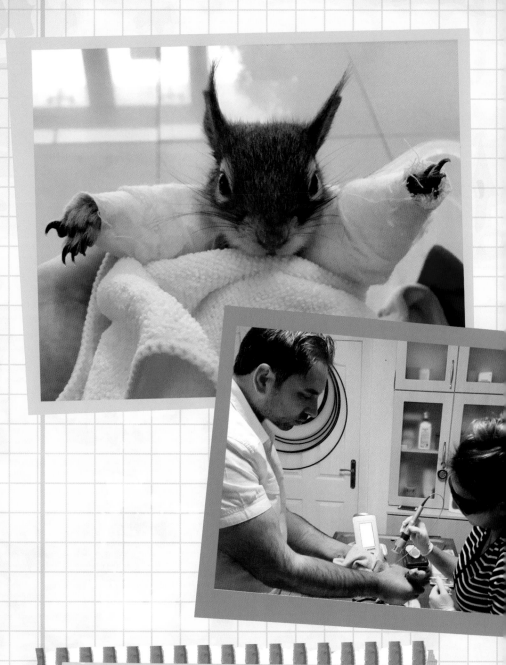

Karamel's rescuer drove seventeen hours to help her.

He took Karamel to a vet clinic right away to get checked out.

Both of Karamel's arms were badly hurt. The vets said Karamel needed to have **surgery**.

They knew removing her arms was the only way for her to get better.

Karamel came through her **surgery** stronger than before.

Soon she began to eat and drink again. Karamel was getting better every day.

She made friends with the other rescued squirrels at her new home.

Karamel even began to run in the garden. But she couldn't run as fast as she used to.

Karamel's owner had an idea to help the little squirrel.

If Karamel had new arms, maybe she could run with her friends again!

So he found someone to make Karamel prosthetic (pross-THEH-tick) arms.

The new device looked like a teeny-tiny wheelbarrow.

When Karamel was hooked in, the squirrel had wheels where her arms used to be.

That made it easy for Karamel to zoom around!

Karamel beat the odds to survive.

Now this little squirrel is living a **BIG** life surrounded by open land and furry friends.

Glossary

alpaca: a South American animal, related to the camel and llama, known for its long neck and silky fur

cleft palate: an opening in the lip or roof of the mouth that a person or animal is born with

clinic: a place where people receive medical treatment or advice

condition: a medical problem that lasts for a long time

device: a piece of equipment that does a particular job

feeding tube: a tube used to feed someone who cannot eat normally

prosthetic: a man-made device that replaces a missing part of the body

ranch: a large farm

surgery: a medical operation performed by a surgeon